Velogen

Humble

PRESENT

to

the

FUTURE

of

ENTIRE MANKIND

If desired, the readers may feel free to contact the author of this book, at the following address.

Guru Prem Walia, P.O.Box 2802

Hamilton, New Jersey-08690-0802, U.S.A.

This book is available all over the world, at most leading bookstores as well as the major retailers on the internet.

On bulk purchase of this book, special discounts are available. For details, please contact the Publisher.

ISBN 978-0-7414-3765-5

Published by:

INFINITY
PUBLISHING.COM

1094 New DeHaven Street, Suite 100
West Conshohocken, PA 19428-2713
Info@buybooksontheweb.com
www.buybooksontheweb.com
Toll-free (877) BUY BOOK
Local Phone (610) 941-9999
Fax (610) 941-9959

Printed in the United States of America
Published March 2007

Velogen

Humble

PRESENT

to

the

FUTURE

of

ENTIRE MANKIND

by
Guru Prem Walia

" The entire known history
of the entire mankind
is the history of
only a few individuals
whose dreams and actions
were different
from the whole lot. "

Shree Swami Vivekanand

the AUTHOR

Like many other authors, I do wish to populate several such pages with lots of degrees, diplomas, certificates, prizes, awards, etc. etc. and etc.

Since the inception of my carried-over memories, I have always been identifying myself as one of the most studious students, aspiring to master the sciences of life.

Several decades have gone bye and I am still a student at the primary level. Quite often my in-abilities have aggravated my in-capabilities and the results have always been uniformly negative.

In spite of countless setbacks and failures, I do consider the time and efforts being employed by me are amply rewarding.

I am learning to notice, to observe, to analyze, and to evaluate the entire history and geography of my own self as well as of those around me. And, in that light, I am learning to decide most of my actions, my re-actions and my in-actions.

I am learning to get access to the infinite ocean of thinking and fully awake dreams.

Also I am learning to choose all my dreams and to crystallize them as best as possible.

Above all, I am learning to over-come all the temptations for all the commercial and financial gains and share the treasure of my most precious dreams with all of those whom I do not even know. This makes me feel closely related to the whole world, without any divisions.

I am learning to learn all this as best as possible. I do hope to go on learning much more.

With lots of hopes for very bright horizons for YOU, for ME and for ALL OF US.

Guru Prem Walia

The ENDEAVOR

PERFECTION in any sphere remains so,
only for the shortest measurable fraction of time.

As the world turns, the realities change and so change the demands of realities, leaving behind an infinite trail of imperfect perfections, which trigger the un-ending quest for more perfections.

This single negative truth governs the continuous process of Human Evolution.

A tiny step forward in the same direction
this is a humble endeavor to address
the two major burning issues
of our present times,

that is,

the ENERGY
&
the POLLUTION.

Velogen

that is how

I would like

to

CALL

or

RECALL

my BABY

my DREAM CHILD

Guru Prem Walia

CONTENTS

01 REFLECTIONS 1

02 OPTIONS 3-28

02.01 Wood 3
02.02 Charcoal 5
02.03 Coal 7
02.04 Fossil Fuel 9
02.05 Bio Fuel 11
02.06 Natural Gas 13
02.07 Solar Energy 16
02.08 Hydrogen Fuel Cells 19
02.09 Hydro Power 21
02.10 Tidal Power 25
02.11 Wind Power 27

03 WIND MILLS 29-35

04 Velogen WIND TURBINES 36-45

04.01 Turbine 37
04.02 Birds Safety Cage 38
04.03 Speed Control 39
04.04 Power Plant 40-44
- A) Gear Box Assy. 40
- B) Coil Springs Assy. 41
- C) Speed Governor 42
- D) Alternator 42-45
  - Disk Rotor 44
  - Disk Stator 45

05 HYBRID DRIVE Vehicles 46-49

06 Velogen HYBRIDS 50-52

07 APPLICATIONS 53-58

08 "CARRY-ON" 59

09 ILLUSTRATIONS 60-67

(Annexure-01)	Ref. pg.	30	on	60
(Annexure-02)	"	30		61
(Annexure-03)	"	31		62
(Annexure-04)	"	35		63
(Annexure-05)	"	42		64
(Annexure-06)	"	44		65
(Annexure-07)	"	43		66
(Annexure-08)	"	43		67

10 Most Useful REFERENCES 68-72

Notes 73-76

0.1 REFLECTIONS

EVOLUTION as known for times immemorial, has been a step-by-step continuous process. The past three centuries have however witnessed this process advancing at unimaginable pace and bringing in all round progress of vast dimensions, in the social, economical and industrial fields.

Advent of steam and electricity facilitated vast expansions of products range. Various new in-puts and the applicable production techniques enhanced the quality controls, automation and efficient productivity, resulting in much larger product-mix and quantitative outputs. Much improved means of communication and transport reduced the distances and extended the horizons of workable markets.

Larger production capacities generated more and more intense competition, forcing the marketing-forces to become more and more aggressive. Innovations started playing the major role in the selection of in-puts, product designs, packing, and the ultimate presentations.

The prime focus remained on enhancing the element of luster and lure in the products, to arrest instant attention of the prospective consumers and trigger their irresistible demand for the product.

Numerous astounding scientific achievements magnified the human stature and practically proved the belief that "God created man in His own image". The sky did not remain the limit any more.

In this head-on blind race, no one has ever had the time or patience to think and realize the ultimate cost of all the laurels mounted on the 'score-board'. Many haphazard plans and actions worked like termites in the foundations of our own gorgeous dream houses. Some of them jeopardized the very fundaments of the most essential ecological balances as provided and maintained by the mother nature.

Some of the industrial in-puts or the production processes emit obnoxious gases and effluents, which are highly poisonous for all human beings as well as other creatures on this planet, including all of those entrepreneurs who are causing it as well as all the consumers for whose pleasures it is caused.

Some of the sources of energy have been consumed without realizing the limitations of their stocks. Some such stocks are already at the rock bottom level. No one cares to think about the fate when the pots would become dry all of a sudden.

In pace with the industrial evolution, the demands are shooting up every moment and the in-puts provided by the mother nature are dwindling faster than ever anticipated.

The pollutants being pumped up in the sky are pulling or pushing the entire world towards the ultimate calamities and the "doom's day".

This bitter truth necessitates immediate steps to take stock of the options available and think of some long lasting remedies.

02 OPTIONS

02.01 Wood

It is the earliest known fuel. Its heat conversion factor is too low to match the modern industrial needs. Hence, its use as fuel is limited mostly to the inhabitants of forests and other remote areas, where no other alternate sources of energy are available. In some urban places, the poverty, the ignorance or just the lethargy are still continuing its use for cooking and domestic heating etc.

Advanced technologies have facilitated development of numerous wood based materials with higher mechanical efficiency and multiples of new end-uses. Pressure treated and seasoned wood, ply-wood, composite boards, particle boards, paper and other wood pulp based products, wood charcoal etc. etc. constitute major end products in bulk quantities.

Trees and plants play the most vital role in the life of all the creatures living in this world. Trees are the prime source of oxygen in the environment, which is essential for breathing to keep alive.
In the natural self balancing process, all the trees and plants consume carbon dioxide as their food, and produce proportionate amount of oxygen for the living creatures.

Hence the loss of each tree or plant results in proportionate amount of imbalance in the natural recycling process as above. In addition, it disrupts the natural habitat and causes ecological imbalances.

Larger amount of carbon dioxide remaining not recycled into oxygen, gets added into the already present amount of carbon dioxide in the atmosphere. Added amount of carbon dioxide causes depletion of ozone layer which releases unfiltered and direct solar radiations on earth, which add to the diseases and deaths. Worst of all, it jacks up global warming with its known fatal results.

Any source of energy, which may further pollute the already over polluted environment, must not even be counted as a probable solution, howsoever easy to obtain and cheaper to use it might be.

02.02 Charcoal

The form of carbon obtained through "destructive distillation" (partial burning or carbonation) of an organic substance is called charcoal.

In addition to charcoal obtained through animal bones, shells of nuts and barks etc., which have a limited use these days, the most widely used variety is the wood charcoal.

Earlier, wood charcoal was obtained by controlling air intake in the kiln for burning stacks of wood. These days it is produced by heating wood in closed rotary furnaces operating at very high temperatures.

While the wood is burnt partially, this method yields acetone, methanol, pyroligneous acid and the acetic acid as bye products, by condensation.

In addition to petty domestic uses, it is widely used as a reducing agent in some metallurgical processes, as a refining agent for sugar, as a filtering agent for gases and some liquid chemicals, for purifying water and for numerous other industrial applications.

The wood charcoal also forms the most important ingredient for making gun powder.

Charcoal out of bamboo is widely used in some drawing crayons and for manufacturing a popular painting ink in Japan, called "Sumi-e".

But as per recent research relating the universal environment, the amount of gases generated through burning charcoal cause much larger amount of global warming than caused by the total quantity of fossil fuel being burnt at present.

Hence, the wood or the wood-charcoal is not what we are looking for as a viable source of energy for the world of today and of the coming tomorrows.

02.03 Coal

It is the world's largest source of energy, made up of carbon, hydrogen, oxygen, nitrogen, sulfur and various amounts of other minerals and moisture. It is mined from the earth's crust.

During earlier stages of industrial evolution, it remained the crowned source of energy. Though much has been phased out by this time, the coal is still being used in very large quantities.

In addition to numerous thermal plants which burn coal to produce steam for serving various industrial requirements or for generating electricity, it is said that even today, more than half of the total world population cooks food and heats their homes by burning coal.

Burning coal produces carbon dioxide, nitrogen oxide and sulfur dioxide. Sulfur dioxide reacts with oxygen and forms sulfur trioxide, which reacts with water and forms sulfuric acid. Solar heat evaporates the sulfuric acid and falls back on earth as acid rain, causing uncountable number of diseases.

Smoke from burning coal includes fly ash, bottom ash, boiler slag and flue gas, which contain arsenic, mercury, lead, nickel, vanadium, beryllium, cadmium, barium, chromium, copper, zinc, molybdenum, selenium and radium. These are the most poisonous pollutants of atmosphere.

Also it contains traces of uranium, thorium and radio-active isotopes, causing numerous radioactive contaminations.

An advanced scientific study claims that the emissions from burning coal for whatever use, directly cause premature human deaths every year, much larger in number than the total number of deaths caused by the entire world war II.

Day by day, the deadly pollutants produced by burning coal in any form, are pushing the universe closer and closer to the point where the extreme imbalances in the scales set up by the mother nature would destroy this planet. That is natural.

Unless and un-till some commercially viable substitutes become available, the vested interests would continue to exert all their might to ensure that this devil dance is not stopped or not even be interrupted for any reason.

God is great but greater are the vested interests of the 'kings-in-process' all over the world.

May God save the God.

02.04 Fossil Fuel

In strict technical terms, the fossil fuel includes coal also. In common language the fossil fuel means the sticky stinking black mud, pumped out from below the crust of earth, which is believed to be the biological remains of large herds of dinosaurs and many other such gigantic pre-historic creatures.

It is processed and refined into large number of consumable products. The biggest part of it is consumed as fuel for automobiles and some other internal combustion engines.

Mother Nature has its own mystical justifications for unequal distribution of this natural treasure on earth. Only a few countries have it as a gift of God. Those who possess it, are the blessed symbols of massive wealth and infinite commercial power. Those who do not have it, often become tempted to grab it by any means, whatsoever.

Over the time, several pages of world history have been written in human blood. Often cropping up temptations and mounting tussles have caused big lots of tragedies, deaths and destruction.

In today's world, the bio-remains of prehistoric dead monsters is claimed to be supporting the lives of human beings on this planet. It seems to be running through the veins and arteries of the entire lot of social, commercial, industrial and political bodies, all over the world.

Ironically, it is one of the biggest culprits that cause the constant depletion of ozone layer, the deadly global warming and the fast growing ecological imbalances, which endanger the very survival of the entire human race and all other creatures on this planet.

In spite of the continuous scientific research for any possible improvements in the production as well as the consumption techniques, the net results so far have not been encouraging. Even the emissions of exhaust gases channeled through any Catalytic Converters are not clean up to the desired levels.

Still, its consumption is accelerating much faster than the imagination. Ever rising demands are nearly licking the bottoms of all known reservoirs.

Whatever the vested interests may claim, proclaim or even disclaim, the horizon seems to be on fire with the signs of RED ALERT.

The entire world is running against time and there is absolutely no scope for complacency.

02.05 Bio Fuel

It is obtained through fermentation and distillation of agricultural products and other bio-degradable elements.

At present, many experiments are being conducted at various levels, to adopt it as an effective substitute for petroleum or may be at least use it as an additive to petrol, to supplement the diminishing stocks of fossil fuel.

It is highly fashionable these days to talk about this new found source of energy. The basic idea appears to be very good, but its ultimate consequences might not come out to be equally good.

It is said that with the available present technology, the crop of soybean grown in one acre of land, would yield nearly sixty gallons of bio fuel, which might fill the tank of just one truck, only once.

Any large scale demand and consumption of this type of fuel would jeopardize the entire pattern of agricultural production all over the globe.

In that case, it would divert the focus on feeding the automobiles instead of growing any food to feed the faster growing human population. Any floods, droughts or inclement weather would aggravate the situation still further.

However, we may hope for the day when our worthy scientists succeed to develop any herbs or plants, suitable exclusively for the production of bio fuel and not fit for the human consumption. It would segregate the resources for production for the human beings and that for the automobiles. It may, to some extent, keep the scales in balance.

It could be something which grows like mushrooms, which are being cultivated indoors, in large trays, in multi-tier installations.

In that situation, we would be free to produce as much of bio fuel as required, by growing any quantity of such raw materials, any where, without ever affecting the food supplies for ever hungry human creatures.

Let's hope.

02.06 Natural Gas

World's largest treasures of Natural Gas exist in Russia, Iran and Qatar. For any large-scale utilization of this source of energy, the shipping, handling, storage and effective distribution have always been the major bottlenecks.

The development of technology for converting it into 'Liquefied Gas' had generated many good hopes. But, for numerous practical reasons, that also could not achieve any considerable commercial break-through.

Some countries have been converting it into agricultural fertilizers. But, its overall cost is higher than other similar raw materials. Hence it has not succeeded to capture any considerable segment of that market also.

The largest market potential existed only as a possible substitute for fossil fuel. The largest consumers could have been the automobiles. But, the vehicles on road are not equipped to run on natural gas or the liquefied natural gas either.

The 'Qatar Petroleum' of Qatar, the 'Sasol' of South Africa as well as a few others are planning to transform the natural gas into liquid fuel, something similar to diesel oil, by "gas-to-liquid" process, as used by Nazis in World War-II, to produce synthetic diesel oil out of coal.

Practically, only the engines based on "compression-ignition" technology are designed to consume diesel as fuel, by virtue of their very high compression ratio. They are employed on very few models of cars and on most of the trucks, buses, tractors, crawlers, marine and some other stationary units.

On the other hand, the engines based on "internal-combustion" technology are not suitable to consume diesel as fuel. This is owing to their much lower compression ratio. They are designed to run only on highly volatile, spark-ignited fuels and not the diesel oil of any description. These engines are employed mostly on air crafts, light transports, most of the cars, motor-bikes, and numerous other stationary and mobile units.

Since these two types of engines are based on two different technologies, it is very cumbersome to convert one type into the other. It involves quite major modifications and lots of expenses.

If in future, any miracle of the modern science proves the GTL diesel oil to be a workable substitute for petroleum and the further models of automobiles come equipped with any new type of engines which are compatible with this new fuel, the cost of converting the existing huge number of engines which are already on the roads or are in the process of being there, from petroleum to diesel oil would be enormous and prohibitive.

Hence, the biggest segment of existing potential market would remain out of reach for diesel oil for quite many years to come.

Any increase in the use of diesel oil as fuel would proportionately increase the amount of particulates, un-burnt carbons, dense smog due to bulk of the nitrogen oxides, obnoxious gases and also it will magnify the ground noise level due to its peculiarly loud knocking sounds. It would deteriorate the already over-polluted environment.

Let's pray and hope for an in-time human access to some un-polluted health resorts on some other planets. WOW !!

Even if at any time the natural gas or liquefied natural gas or the new type of diesel oil succeeds to become the widely acceptable substitute for petroleum etc., all of its known stocks and reservoirs would not suffice to last long enough to meet even a part of the total projected global demand.

Sooner than soon, the world would still be starving for any long term dependable sources of energy.

02.07 Solar Energy

Amongst all other sources and forms of energy available on this planet, the Solar Energy was the first amongst all others, recognized by the human beings.

Within the known history of this universe, from the times immemorial, the Sun has been worshipped all over the world. The Egyptians considered Sun as their first king who formed Egypt and for them their king was their God. In Greece, the Apollo as well as the Helios were considered the two forms of the Sun God. Even today, Sun is worshiped by many, by one or the other name.

Even those who do not believe in God and religion, do understand that Sun is the prime source of energy that supports the very existence of all the creatures big and small, all the plants and vegetation and all the known or unknown forms of life on this planet.

So far, the Solar energy has been availed in its most prime forms, as it comes. In today's world of human needs and the relevant scientific advances, it has become essential to try to make use of all available forms of all the sources of energy as much as possible.

During late 18th century, the Solar Water Heaters were developed in USA and in other parts of the industrialized world.

These heaters were a big leap forward towards the direction of efficient use of the available energies. Instead of the usual wood or coal burning ones, these heaters used sunshine. By early 19th century, big lots of these heaters were in vogue in most of the cities.

With the advent of electricity in late 18th century, the quest for producing electricity by means of solar energy took birth. The shape and size of solar water heaters were improved to produce steam by heating water in large solar power plants. The steam thus produced was employed to drive the turbines which drove the electric generators of large capacity to meet the commercial demands

Earlier such plants used a highly curved mirror, called the "parabolic trough" which focused the sun rays on a sequence of water bearing pipes. The solar heat thus concentrated, turned the water into steam.

Another form of solar power plants to produce electricity was called "Heliostat". It comprised of a large number of mirrors in panels which moved and turned to face the sun all day long. These panels surrounded a suitable size water tower. The sun rays focused on the water tower which produced steam and thus the electricity.

In the present times, the development of photo voltaic cells to produce electricity through solar energy has been a very bright idea. Such cells were developed in 1950s for use on US space satellites. They were made of silicon.

In addition to their use on space shuttles and the satellites, they have been used on a few models of experimental cars, a number of small or medium size gadgets or on the rooftops, for domestic heating as well as a few more such needs.

Their ultimate commercial viability depends upon the continuity of their functioning at all the places, all around the year. It depends entirely upon definite availability of sunshine, which is just not probable.

Secondly, their high initial cost is keeping them beyond the reach of a common man.

In context of the overall global requirements of energy, at present one can only hope for more promising developments of commercially viable technology.

All human hopes bank upon the God's Will and some-one's skill.

02.08 Hydrogen Fuel Cells

Way back in 1839, a scientist named William Grove developed a simple acid fuel cell, by putting some electrolyte liquid in a container, with two electrodes, one of them a negative anode and the other one a positive cathode.

He arranged to supply hydrogen to the anode and kept the cathode in contact with the air. The electrolyte caused hydrogen on the anode side to be ionized, thus separating electrons from protons.

The ionization process thus released the electrons, which when collected, produced electricity. Also it released protons as hydrogen ions.

The electrolyte prevented the free electrons from reaching directly towards the oxygen, which was inviting them at the cathode.

That made it essential for the electrons to travel towards protons only through any external passage. That passage was the wire that carried electricity.

Beyond that stage, within these nearly two centuries, the scientific acumen has taken numerous flights towards more and more perfections.

There are five major types of fuel cells, produced through five different processes, each one having different properties, different costs, different ways of applications and radically different pros and cons.

1) Proton exchange membrane cells.
2) Alkaline fuel cells.
3) Phosphoric acid fuel cells.
4) Molten carbonate fuel cells.
5) Solid oxide fuel cells.

Some varieties of hydrogen fuel cells have already been used successfully on rockets, sub marines and sports cars.

For any steps further, the differences of opinion at the topmost level raised the hedges. The technical feasibility, the long term reliability, the commercial viability, etc. etc. and etc. overcame the ability to agree upon any of the viable options.

When all the Gods agree unanimously, not to agree upon any matter of universal applications, the victim is always the one like hydrogen fuel cells.

The only saviors could be some plural number of commercially potent Godfathers, with keen competing interest to see the venture grow and be productive as early as possible.

Only they could ensure that the baby is not labeled "still born" and is not buried alive, without ever getting a head stone for the grave.

To reach the launching pads, the whole concept seems to be waiting for a larger combination of factors, which do not seem probable at present.

At present, the only option available is, to keep our hopes alive for the future of Hydrogen Fuel Cell and also remain on the lookout for other possible sources of energy.

02.09 Hydro Power

Thrust of flowing water is employed to rotate water turbines which drive the generators to produce electricity.

The required amount of thrust may be created, regulated or boosted through water dams, water tunnels, channels, pumped storage reservoirs, or natural waterfalls and streams etc.

In addition to the thrust of flowing water, the kinetic energy of Tidal Waves has also been used to produce electricity. It is being discussed separately in the following chapter.

Hydro Power is a renewable source of energy, with very low recurring cost of production. Multiuse Dams serve for irrigation, flood controls, tourism, recreation, water sports etc. Plants for Hydro Power are more durable than any other power plants. Some plants which are in regular service were built over a century ago.

The earliest known Hydro Power plant was installed in Rothbury, England, in 1870.

The earliest known Water-wheel in USA was built with Morgan Smith Turbine, in 1869, at Stottville, New York State.

First commercial supply of Hydro Power started in 1882, from Appleton, Wisconsin, USA. The second one in USA was in Minneapolis in the same year.

First publicly owned Hydro Power plant was built in Tasmania, Australia, in 1895.

First one in Canada was built in 1898, on Decew Falls, Ontario.

The world's largest Hydro Power plant is located in Quebec, Canada, with over 16,000 Mega Watts production capacity.

Larger than this is now under construction in China, designed to produce over 18,000 Mega Watts of electricity and is expected to be ready for production in the year 2009.

<u>Major global INSTALLED CAPACITIES</u>

USA	79,511 MW
Canada	66,954 "
China	65,000 "
Brazil	57,517 "
Russia	46,100 "
Norway	27,528 "
Japan	27,229 "
France	25,335 "
India	22,083 "

During heavy rains, the flow of water from the Dams has to be reduced to avoid flooding. Reduced flow of water reduces the production of Electricity. Most often this jeopardizes the constant flow of electric supply and causes all the resultant problems.

Large Hydro Power plants disrupt the natural ecological balance of that area, up as well as down-stream, specifically affecting the fish and fauna.

The water coming through Turbines has very little suspended sediments which are necessary for maintaining the river banks as well as the river bed from scouring. So, the down-stream river is affected adversely.

Since the quantum of water flowing out of Turbines is frequently regulated to match the demand of electricity, it causes considerable surges in the water flowing down the river. It causes erosion in the sand bars which are essential to control the sudden change of flowing direction of the river.

Since all the human estimates are not always right, some hydro power plants and their huge water reservoirs are located by mistake in some unsuitable geological areas. It causes local earthquakes and unbalanced rainfalls as well as floods due to disrupted rainwater drainage systems.

In similar extreme conditions, such projects cause long spans of totally dry river beds, which affect regular water supply to the centuries old inhabitants on river banks, their crops, their cattle and their entire system for life support.

To build each hydro power project it becomes essential to dislodge quite large population, with their centuries old attachments and affinities for those locations which are deeply embedded with their ancestral traditions, their culture and all sorts of memories. It disrupts the entire social fiber of that area and causes long lasting resentments. Any amount of compensations can not heal it for many generations to come.

Above all, various green plants and vegetations covered by the water stored in reservoirs to run the hydro power plants constantly decay and produce substantial amount of methane gas and carbon dioxide, which are the major culprits for global warming and the greenhouse effects.

In some cases, such methane gas emissions are larger than the quantity emitted through many electric generation plants using thermal power.

The known scope for any ambitious additions or expansions to the existing hydro power plants seems to have already crossed the saturation point.

Hence, continued hunting for other alternate sources is most essential.

02.10 Tidal Power

The Tides move intermittently, some enormous amounts of water-mass from one direction to the other. This results in occurrence of the following two types of potential energy force.

i) Kinetic energy created by the tidal waves, owing to their surging and ebbing actions.

ii) Force of the water-mass, rising high with the tide and falling with the ebb. Level of height it reaches is called the "Head".

To avail the said kinetic energy, various on-shore Dams are built with their height proportionate to anticipated amplitude of the incoming tidal waves.

Suitable size river-like water channels are linked with the sea. The high tides funnel water through those channels into narrow passages fitted with suitable submersible water turbines.

The ebbing waves return through similar passage and thus drive the return-turbines. All of these turbines drive generators and produce electricity.

It is considered less expensive, more feasible and more popular than using the said "Heads" of the water-mass. Tide mills have been used in Europe for over 1000 years, mainly for grinding corn.

The only important factor is the correct choice of their location. Amplitude of tidal waves determines the capacity to produce electricity. Tidal power is largely predictable.

To avail the "Heads" of the water-mass for producing electricity, very huge barrages are built within the ocean, to trap water of the high tide.

For this, several pre-caste concrete blocks of very large size, called the 'caissons' are placed in position to build a wall or the dam for the barrage. Very long and large embankments are built from one end of the caissons, around the periphery of the barrage, to the other end of the caissons. This virtually builds a deep barrage, within the deep ocean. The caissons hold a 'ship-lock', 'sluice- gates' and two or more of submersible water turbines.

Initially water collects in the barrage through the sluice gates which are open till the high tide comes in. Thrust of the incoming water rotates the submersible water turbines to produce electricity.

As soon as the high tide comes in, the sluice gates are closed and maximum amount of water raised-up by the high tide (head) is trapped within the barrage by the time the tide ebbs down.

Thrust of water at height (head) in the barrage turns the turbines in the fixed in caissons and they drive the coupled generators to produce electricity.

This process goes on in the deep ocean and the electricity produced is cabled up to the shore and then transmitted and distributed as required. This method also could not succeed much, as it is too cumbersome and expensive and is comparatively less productive. Hence, not the required solution.

02.11 Wind Power

Of all the known sources of energy, wind power seems to be the prime hope of heaven.

It is abundantly available all over the globe and by using it to generate electrical energy, its stocks are not going to be "consumed" in any sense.

It is free of cost and well within the reach of any one, at any time, any where on this planet.

The point to ponder is, that this omni-present powerful energy would have certainly attracted the attention of even the earliest mature human mind. It is equally obvious that some of them must have strived to avail this energy, much before their attempts on any other such source.

In the medieval ages, they started making use of the wind energy to propel the 'sails' of various sizes and shapes of boats and small ships.

As late as first century A.D., some unknown wizard in some remote village of Northern part of China had conceived and successfully employed the wind energy to push-drive the wind-sails mounted on rafters connecting an axle, thus obtaining a pivoted rotary motion to the axle. On the axle was mounted a grinding stone, with its counter-part placed on the ground. This was perhaps the first known wind mill, conceived, designed and successfully employed to grind corn etc.

Since then, there must have been some valid reasons for the inability or failure on the part of the scientific thinkers all over the world, to think and proceed even a few steps beyond that.

Hence, it can not be considered as a readily available piece of cake. There must have been some valid reasons for the absolute failure of one and all.

Before riding any wishful anticipations, it seems imperative to dig-out the bare facts, and try to analyze them as far as possible and then think of all probable solutions.

03 WIND MILLS

Concept of availing the energy of wind for industrial use was born somewhere in North China, over 2000 years ago.

They succeeded to use the energy of blowing wind to push the cloth masts which rotated the wind mill on its vertical axis. At the lower end of the axle they mounted the grinding stone.

So, instead of any men or animals, the wind started driving the grinding stone to grind the grains. They found it more convenient and much more economical.

The only authentic support to prove it is found in their local folk songs, in praise of some unknown wizard who invented it.

Owing to lack of communications it could not travel much beyond its place of birth for over 700 years.

Sometimes between 700 and 900 A.D. some mill-smiths from North China traveled along a silk caravan to Persia.

There, the demonstration of their unique skills was so tempting, it is said, the rulers decided to treasure it solely for their own clan. So, they made those mill-smiths as their life time 'guests', living within their given premises in Persia.

Thus, the wind mills were designed to operate behind walled enclosures, with small openings in a corner for the air intake. The know-how remained arrested behind those walls for over 500 years.

(Annexure-01, page 60)

Some Persian folk songs narrate that sometimes around 1200 A.D. a Dutch family was on a pleasure trip to Persia. One of their glamorous young members attracted the trust of a Persian youth, who was successor to the custodians of a wind mill in that area.

The young Persian invited the guests behind those walls, to witness the boastful demonstration of their unique family assets. Thus the age old secret technology for wind mills was set free from the enclosed walls in Persia, to the open fields in Netherland.

In Netherlands, they started with the basic Persian design and installed them in their often windy open fields. They used these wonderful machines, for grinding corn and for pumping out water from their often flooded drains.

(Annexure-02, page 61)

Very soon these new found "wonders" attracted hot public discussions and loud praise all over the country. Their government as well as the financial institutions provided all out support for research and design development, which gave birth to the legendry Dutch Wind Mills for numerous fruitful applications.

These were employed for grinding corn, barley, mustard, cocoa, pepper, lime, chalk and for hulling rice, sawing timber, and for operating their textiles mills etc.

By the end of 18th century, they had more than 9000 wind mills working all over Netherlands. Western De Zaanse region had the largest number of them. About one thousand still exist there, as "Protected National Monuments".

From Netherlands, this technology flew freely and quickly to Europe and other parts of the world. The "Pilgrims" carried it along to USA and Canada.

There is an interesting part of this episode. It is said that America's "Declaration of Independence" in 1776, was written on a parchment from a wind mill named “De- Schoolmeestr”, of "De Zaanse Schans" region in Western Netherlands.

Birth of electricity in 1831 and the successful development of the steam engines within the same decade, killed the status of wind mills every where. Most of them were closed down.

In USA and Canada, some entrepreneurs developed the wind turbines. The age old sails made of wood or reef were replaced with more streamlined impeller blades of metal, rotating on horizontal axis. They employed the wind turbines not for grinding grains, but to drive electric generators, for producing electricity.

(Annexure-03, page 62)

From USA and Canada, the newly evolved technology for wind turbines flew very fast to Europe and other parts of the world.

They further developed many new models, sizes and shapes, to suit their requirements. They installed them in open fields, on hill slopes, and on sea shores, in large numbers and called them the "Wind Farms" and "Energy Parks".

In USA, the centuries old philosophy of only the larger-the-better, prevailed in this field also. The promoters used all their means to bring out the wind turbines which could be boasted about as the biggest ever built on earth.

In 1930's, Palmer Smith Putnam of USA designed 1.25 Mega Watts capacity wind turbine, with two impeller blades made out of stainless steel, having 175 feet wide rotor span, weighing over 16 Tons, rotating on horizontal axis, at 28 rpm.

It was mounted on a steel structure which was taller than a twenty storey high building. It is said to be actually the largest wind turbine ever built.

By mid 1941, it was installed and fully operative in Vermont. After remaining in use for a very short time, one of its blades broke off near the hub, for real reasons still not known.

Overall demand for electricity was visibly riding rockets everywhere. Some of the entrepreneurs in USA and else where had launched a number of very ambitious projects for production of electricity by using thermal or hydro energy. Their obvious success promised each one of them a lion's share of the golden catch.

On the other hand, the wind turbines were facing rough weather. Though the main input was amply

available free of cost, the major problem was to ensure uninterrupted flow of electricity during low wind velocity, when the wind turbines could not rotate at the required speed. The very thought of any such black-outs seemed repulsive.

Second strong objection against the wind turbines was raised by the environment protection organizations in many parts of the world. They insisted on avoiding the risk of wind turbines hurting or killing any flying birds.

In Netherlands, mighty lobbies of environmentalists got it officially proclaimed that all the wind energy parks near any inhabited areas were detrimental to the public health and even at other locations, they were awfully risky for the birds and also very harmful for the sea life.

Exactly similar situations arose in U.K., Germany, Australia, and many other parts of the world.

The real major factor every where was the direct clash of interests between the promoters of wind turbines and the producers of thermal and the hydro generated electricity.

By then, the opponents of wind turbines had secured ample resources to win all-out support from the rulers, the financers as well as the general public. In spite of all the desperate efforts by the promoters of wind turbines, their opposing vested interests remained victorious for a very long time.

All further plans and projects for the wind turbines got jacked up at some unknown wishful stations. The wind turbines had virtually gone with the wind.

Way down, during 1974 and 1981, the U.S. Federal Energy Plans had injected a new life into the wind turbines. The innovations appeared to have started breathing again.

Development of vertical-axis turbines with straight-blades, cyclo-turbines and the gyro-mills etc. brought in lots of new hopes. Development of fiber glass blades instead of metals was a big step.

But, as the world turned, the realities changed and so changed the demands of realities. In 1981, the new rulers brought new rules, which reversed the process of recovery. Perhaps it was God's will.

Above all, the biggest possible plans, for the biggest in size wind turbines, built to form the biggest possible energy parks, with the biggest possible transmission and distribution network, embodied much bigger lots of inherent, unforeseen, technical, commercial and management problems.

This has been the biggest factor, that crushed down most of the perfectly conceived plans under the dead-weight of their own "size-oriented-problems".

In the present situation, the real success lies in re-thinking and re-modeling the entire concept.

In addition to the production of ten giant size wind turbines, it seems more appropriate to plan and produce simultaneously, at least ten thousand units of medium and small size wind turbines, like the size of a pedestal fan or even smaller, to serve all the needs of an individual consumers, any where and every where, as per details in next chapter.

04 Velogen Wind Turbines

So far, all the wind turbines have been planned and designed with entire focus on the size, which was the biggest of all, meant to be installed in the largest possible energy park.

They have impeller blades which look like the ones of an airplane, rotating on horizontal axis. The generator is mounted on tail-end of the axle. The whole apparatus is mounted on some high altitude fixed tower or a pillar.

The Velogen Turbines would be based upon absolutely different approach to the whole concept of wind energy. They would have numerous shapes, sizes as well as their working mechanisms. Each one would be designed to serve the specific needs of each application.

Most of the Velogen Turbines would look like a Torus or a Mushroom, rotating on its vertical axis, comprising of the main assemblies, as per details discussed below.

(Annexure-04, page 63)

04.01 Turbine

Between the upper and lower rims of the turbine, it would have a number of air-pockets, forming small webs, to peddle the turbine.

In addition to said web of air-pockets, there would be another number of telescopic air-ports, designed to change automatically their size and direction, to regulate speed of the turbine.

In addition to the above air-pockets and air-ports, it would have a number of "swirling wind tunnels" with wider mouth on the outer rim, "coning-in" towards the open corridor surrounding the axle of the turbine. It would boost the torque.

(ANNEXURE-04, page 63)

04.02 Birds Safety Cage

The entire turbine would be encaged in suitable wire-mesh grill, to keep all sorts of flying birds safely away from the turbine blades and thus make sure that they are not sucked-in, injured or killed.

This security device would surely satisfy the prime concern of all the bird lovers and environmentalists all over the world.

(ANNEXURE-04, page 63)

04.03 Speed Control

At immediate next lower level of the turbine housing pedestal, there would be a conventional brake drum, with couple of fly-weights type brake-pads mounted on the axle.

The centrifugal forces of the rotating axle would push the brake-pads outwards, to rub the wall of the brake drum.

This would automatically reduce the speed of the turbine and provide automatic safe control of the possible "speed-over-runs" due to un-usually high wind velocity at any time.

(ANNEXURE-04, page 63)

04.04 Power Plant

At next lower level on the housing, the main Power-Plant would be mounted. It would have the following major Assemblies, as per details discussed below. (ANNEXURE-04, page 63)

A) The Gear Box Assembly.
B) The Coil Springs Assembly.
C) The Speed Governor Assembly.
D) The Velogen Alternator Assembly.
   Velogen Disk Rotor.
   Velogen Disk Stator.

A) Gear Box Assembly

The Gear Box would be specially designed, to automatically perform the following three main tasks, independent of each other.

i) When sufficient Wind Velocity is available.

To route the power from wind turbine axle, to the coil springs and thus screw-wind them up to their full capacity and then lodge them in the "stand-bye" position, ready to recoil as and when needed.
Next,
to leave the coil springs alone and divert the power from the wind turbine axle directly to speed governor and drive the disk rotor to produce electricity.

ii) **When sufficient Wind Velocity is not available.**

To disconnect the direct link between the disk rotor and the wind turbine axle and connect the fully wound coil-springs to the speed governor, on to the disk rotor.
Hence,
while the turbine is idle,
the coil-springs would recoil and
keep the disk rotating,
and ensure uninterrupted supply of Electricity.

iii) **When sufficient Wind Velocity is available again.**

To repeat the operations as per (i) above.

B) Coil Springs Assembly

It would have a number of very high tension Coil Springs, connected to each other in series, geared to carry out the functions, as per above and keep the Disk Rotor rotating, without any break, even when sufficient Wind Velocity is not available and the Wind Turbine is idling.

IN ADDITION TO ABOVE,
the medium, small and miniature size Velogens would have the provision for winding their coil-springs, by HAND-CRANKING, like the clocks.

It would enable fully localized and independent production and supply of electricity, even in the most remote places like underground mines etc., without any need for the wind power or any external source of electric supply.

C) Speed Governor Assembly

It would get its in-put drive through the gear box, either directly from the turbine axle or from the coil springs assembly.

The main in-put drive would spin a number of fly-weights, each one mounted on a steel strip, with their one end fixed on a small size flywheel, which would be sliding on a turret base, responding to the dynamic position of the spinning fly-weights, varying according to the speed.

An adjustable soft brake-pad applying on the rim of the flywheel, would govern speed of the flywheel and determine speed of out-put spindle. The out-put spindle would drive the disk rotor, at the speed, as determined through positioning the soft brake pad. Thus, it would govern the out-put of Velogen Alternator.

It is something similar to the Speed Governors, mostly "governing" the playing speed of the Grand Pa's age-old Gramophone Machine.

D) Alternator Assembly

Looking at the existing designs of the machines for producing electricity, the ideal objectives for further design developments would be, as below.

a) Power in-put needed for start-up should be minimized.

b) Workable Current out-put should be maintained even at lower revolutions.
c) Production output per revolution must be maximized.
d) Voltage be accurately regulated.
e) Cost of manufacture and of operation as well as maintenance, to be the lowest.
f) Uniform un-interrupted out-put must be ensured, irrespective of Wind Velocity.

The conventional generators and alternators mostly have two main components,
i.e. the Rotor and the Stator, as discussed below.

The Rotors in vogue so far, are of cylindrical shape, like common kitchen-roller-pins.
Both of its ends are supported by bush-bearings and or ball-bearings. One end of the Rotor-shaft has provision for external drive. Further details about the design and positioning of its 'collector-coils' depend upon its output,
i.e. A.C. or D.C.

The Stators so far, are mounted on the inner shell of the casing. There are a number of magnets, surrounded by winding wires, called the 'field coils,' with regulated power supply, to govern the magnetic flux.

To serve all the objectives defined above, the A.C. Alternators as per details described below, appear to be more versatile. The reasoning for this conclusion may be too long to avoid deviation from the main topic, hence ignored.

Velogen DISK ROTOR

Instead of a cylindrical rotor, it would be of flat round disk shape, of various diameters and thickness, as per details described below.

(Annexure-05, page 64)

a) It would have a number of permanent magnets, surrounded by field coils, connecting the external voltage regulator, through two insulated Slip-Rings, mounted on the lower radius of the Disk.

b) The Disk would rotate on a double ended Axle, with "taper-pin-point" ends, resting and floating on some hard jewels or some industrial diamonds at both ends.
(No Bushes or Bearings)
This would minimize the requirement for start-up and driving in-puts and maximize the acceleration and speed.

c) The double sided Disk Rotors with double Stator Disks would maximize the output per each revolution.

(Annexure 07 & 08, pages 66 & 67)

d) The Spindle driving the Rotor, would have helical grooves, driven through either of the following two sources:-

i) WIND TURBINE > COIL SPRINGS > GOVERNOR > DISK ROTOR.
(when sufficient wind velocity is available).

ii) COIL SPRINGS > GOVERNOR > DISK ROTOR.
(when sufficient wind velocity is not available)

This would ensure un-interrupted production of electricity, without ever having to depend upon the availability of 'sufficient' wind velocity.

Velogen DISK STATOR

It would be of disk shape, suitable to mate with shape and size of the Rotor.

It would have "Collector Coil Windings" around a number of Core Segments, matching the Magnets on the Rotor.

(Annexure-06, page 65)

The Magnetic Flux of each Field Coil on the Disk Rotor would be interrupted by the conductors of Collector Coils on the Core Segments of the Disk Stator.

It would create electric impulses which would be collected by the Collector Coils and passed on to the Out-put Terminals of Alternator, as Electric Current.

All of the above points combined, would hopefully bring out the Alternators of desired efficiency.

05 Hybrid DriveVehicles

A hybrid drive vehicle (HV) is the one using a combination of energy inputs for its traction. Most popular form is the installation of re-chargeable energy storage system (RESS) on board and a smaller internal combustion engine, powered by any of the conventional fuels. Since most of them use petroleum based fuels, these Hybrid Vehicles are known as petroleum-electric-hybrid vehicles (HEV).

Various permutation and substitution methods are being used to determine the best combination of hybrid to ensure the most fruitful use of these power sources.

In most of the cases, the start-up is powered by the batteries on board, which drive electric motors mounted on each wheel. As the speed gathers momentum, the load on batteries is gradually shared by the electricity produced by a small generator driven by the internal combustion engine on board. In addition to driving the traction motors, the small engine driven generator operates to re-charge the batteries.

Such vehicles consume much less fuel and can attain the acceleration and speed like any gas consuming one. It emits much less amount of pollutants.

Similar to the HEVs, there are the 'all electric' vehicles, which can run on batteries, charged by external sources of electricity for this purpose.

A small number of experimental vehicles are equipped to run on hydrogen fuel cells.

Almost all the hybrid vehicles still require some quantity of gasoline, diesel or any other such fuel.

The hybrid vehicles are not any recent concept. They have an age old history of development.

In 1898, Porsche produced a 'series-hybrid vehicle', called "Lohner-Porsche-Carriage". In the year 1901, Mr. Ferdinand Porsche himself drove one such car and won the Exelberg rally. In addition, it broke several Austrian speed records. Over 300 of these were sold to public.

In 1915, the Woods Motor Vehicles produced a car called 'Dual Power', with a four cylinders internal combustion engine and a battery powered motor.

Below 15 mph speed, the battery operated electric motor alone drove the vehicle and above that, the engine took over to drive it up to 35 mph top speed. By 1918, about 600 of them were produced and sold.

In 1959, National Union Electric Company, Henney Coachworks, Renault and Eureka Williams Co. joined hands to produce the first electric car, with electronic speed controls. Such cars were called "Henney Kilowatt". These cars had paved the way for modern electric and hybrid cars.

Between 1960-70, a scientist named Victor Wouk, who was previously employed by 'Henney Kilowatt', developed a few more advanced models, of the HEVs, for which he was commonly called "Godfather of the Hybrid".

By 1978, an electrical engineer named David Arthur developed a DC generator and a voltage controller that linked the batteries. That enabled the HEVs to cover 75 miles per gallon of fuel. This concept was called "Regenerative-Braking-Hybrid". It increased the life of charge on the batteries. It was a vital step forward in the development of the modern Hybrids.

Since 1990, the successful commercial application of hybrid technology on automobiles took strength from the HONDA-Insight and TOYOTA-Prius.

Now there are quite a large number of such or similar vehicles developed by many other prominent automobile manufacturers all over the world. Each year, the new models come up with more and more advancements and plus points.

In the year 2005, the Ford Motor Company brought-into the market, the first hybrid powered sports utility vehicle, named "Ford Escape".

Toyota introduced their "Highlander", Lexus came out with their "RX-400-h" and "GS-450-h", Nissan offered "Altima-hybrid", Honda "Civic-hybrid" and Honda "Accord-hybrid".

In addition to the automobiles, similar hybrid technology is being widely used on Diesel-Electric-Locomotives and some models of Sub Marines. Lately, some prominent manufacturers are coming out with their models of hybrid busses and trucks.

Obviously it is the fruit of toil and concerted efforts of a large number of scientists and engineers all over the world. Whatever has been achieved so far, deserves loud appreciations.

But, irrespective of the capacity and the efficiency of batteries installed to drive any such automobiles, there would still be the need for some quantity of gas. And, gas would not be available for long.

In addition to the hybrids mentioned above, the Velogen Hybrids comprising of a combination of wind power and the electric batteries holds bright promises. The details of this concept are discussed in the following chapter.

06 Velogen HYBRIDS.

The most useful application of Velogen Wind Turbines would be for the automobiles of all sizes, shapes and descriptions. Some basic details of the concept of Velogen Hybrids are discussed below.

Anything, of any shape or size, when in motion in any direction, at any speed or momentum, it does have to face the proportionate amount of resistance, forced by the presence of air, exerting itself to stop that motion.

Hence, while designing the body or power train of every Automobile, the Wind has always been considered as the Negative Force.

Application of Velogen Wind Turbines on the Automobiles would surely be unique and many steps ahead of all others.

As explained below, the usual Negative Force would be fruitfully availed to provide the required driving force to the Automobiles, pollution free and absolutely free of cost whatsoever.

In other words, Velogen Automobiles would go for over a million miles per gallon, with much better speed and acceleration.

At the face of it, it sounds like a wishful thinking, a dream. The fact remains that all the amazing human achievements, ever since, till date, have been the legitimate off-springs of some one's dreams; some one, mostly unknown.

Of course, for such concepts, the final details of the picture emerge only after going through the complete cycle of development, from ideas to imaginations to drawing tables, to assembly lines, to test benches, to the endurance tests and finally, the consumer's response.

To qualify for ingress into any such cycle of developments, here are some details.

1. Just like any other Hybrid Car, it would require an Electric Battery of workable capacity, mounted on the chassis.

2. In the Bonnet Chamber, immediately facing the front Grill, there would be a number of "Rowing-fins" type cascading Wind Turbines, on cross-horizontal axis, to optimize use of the incoming flow of resisting air stream.

3. These Wind Turbines would drive a number of suitable size Velogen Alternator assemblies, coupled to the Wind Turbines, in the Bonnet Chamber.

4. To drive the Vehicle, the Electric Battery would directly crank the Traction Motors, mounted on each wheel. Thus, the Vehicle would initially move at the speed, as determined through the Accelerator.

5. Any movement of the vehicle would generate the proportionate amount of air resistance, opposing the motion of the vehicle.

6. Unlike conventional Generators or Alternators, the design of Velogen Alternators would require just a tiny fraction of the energy to start and to keep the Alternator Disks rotating at the desired speed.

7. Amount of electricity produced by Alternators would be proportionate to the amount and force of the in coming flow of air.

8. This flow would be optimized through some "Swirling-Wind-Tunnels" of appropriate shapes and sizes, surrounding the Wind Turbines inside the Bonnet Chamber.

9. Even at city driving speed limits, the out-put of electricity can reasonably be anticipated to prove sufficient for sharing gradually and taking over ultimately, the entire Work-Load of Battery and also keep it fully re-charged.

10. Thus, the actual total cost of running per mile would be, merely the cost of periodical repairs and maintenance of the whole system.

Once the basic details of the entire concept reach the minds of those, who have the capability as well as capacity to take it further, the ultimate reality in driving such automobiles, would not, surely would not remain a distant dream.

07 APPLICATIONS

* The giant size Velogen wind turbines would be planted on the sea-shores, river-banks, hill-slopes, open fields, valleys and plains etc. in appropriate configurations, to constitute a chain of energy parks, inter-linked through suitable transmission loops and grids network, primarily to serve the bulk consumers.

* Combinations of giant and medium size wind turbines would prove to be the most useful on decks of ships, tankers, container carriers, cruise-boats, or anything sailing amidst flow of air of any velocity, from any direction.

* For some of the vessels, it would serve their requirements for lighting and other utilities and for a few others, the planned installations of wind turbines would drive their traction motors.

* Adequately planned and properly located wind turbines on the roofs or the chassis of railway trains would surely save billions worth cost of the conventional energy in-puts.

* Appropriate size Velogen wind turbines would be an asset for various Drilling Riggs and other Platforms, on-shore or off-shore, or at high seas.

* They would serve equally well at various Installations, Establishments and Observation

Centers in remote areas, where conventional channels of transmission and distribution are either not available or not feasible due to reasons whatsoever.

* Medium and small size portable units or fixed-base Velogens of suitable designs, driven by wind or tiny hydro power, or even hand-cranked coil-springs, would surely perform like 'Genies', to serve various requirements in the most remote areas, where electricity could not reach so far.

* <u>These GENIES WOULD</u> ....

  a) light-up their huts, their barns, and their paths.
  b) energize their cooking-ovens, their heaters, or their coolers.
  c) fetch water for their daily needs, to their own door-steps.
  d) irrigate their small fields, crops and their cultivations.
  e) operate their farm equipments, cutters, thrashers etc. for better agricultural yields.
  f) serve their schools, dispensaries and their community centers.
  g) promote cottage industries, artisan shops and handicrafts.
  h) promote small food processing and preservation units
  i) create abundant self-employment as well as job opportunities
  j) enhance overall human productivity and all round prosperity.
  k) keep them 'related' to the 'other world'.

* The cities and suburbs hold equally bright promises for some light weight, fixed or portable, medium and small Velogens operating on wind as well as on hand cranked coil springs, placed in their windows, their balconies, court-yards, back-yards, their lawns, drive-ways, or their roof-tops, like dish antennas.

* All that one has to do is to place it or shift it to any corner of the house, where comparatively "more" of wind velocity may be probable.

* Keep their coil springs fully wound by hand cranking, always in the 'stand-bye' position.

* Further, they have to just plug the Velogen output cord into any of the domestic wiring wall-sockets and run happily ever after, their domestic lights and all the gadgets, without ever having to pay any bills for their consumption of electricity.

* Initial purchasing cost is the only investment they would have to make and the bills for periodical repairs and maintenance of the Velogens is the only recurring expense that one would have to incur.

* The Velogens would be very effective on river-banks, canals, streams, springs, water falls etc. Small persian wheels or suitable size water-turbines would drive the Alternators.

* There is enormous scope for small and miniature size Velogens also. The Miners, Surveyors, Rangers, Hunters, Hikers, Campers, Guards,

Border-Patrols, Foresters, Mountaineers, plus many more, would always need some sort of independent source of electricity for their numerous essential requirements.

* A Velogen with Disk Rotor of only six inches diameter, when mounted on the top of a tent, would light a lamp and also warm up the blanket, the bed as well as the entire tent. Or, drive a small air-conditioner or a cooler in a burning hot desert.

* A small handy Velogen, operating on wind or any water stream, would be able to generate sufficient power to operate a chain-saw, to fell the trees or to cut logs in deep forests.

* A tiny Velogen, with Disk Rotor of less than three inches diameter, when mounted on an elastic band and worn like a cap, or a band around the bicep, or tucked into any button hole, can surely keep the interior of a snow suit warm enough, even in Alaska.

* Velogen Turbines and Alternators would make it possible to save the huge investments in long transmission lines, distribution networks, grids, transformers, and their costly and often painful maintenance in hostile conditions.

* So far, there is little human control over natural calamities and their disastrous disruptions. The earthquakes, tsunami, cyclones, storms or the hurricanes may occur without least expectations.

But, come what may, even the small and medium Velogen Alternators would keep functioning with the wind power or by hand cranking the coil springs if needed. This would definitely keep all the essential services intact and un-interrupted under all circumstances.

* Major achievements of the Velogen would be, to localize the production and supply of electricity, in each and every sphere of its actual demand. It would abolish the role of various intermediaries and via medias.

* It would liberate all the consumers all over the world, from their recurring liabilities for the un-limited "Conventional Management Costs", the huge amounts regularly borne by consumers and consumed by the overheads, mostly leaving behind some "simply complex" statistics for the public consumption.

* In face of the howling realities about the future of most of the existing sources of energy particularly the Fossil Fuel, some radically new realignments of market forces can not be ruled out. Let us be ready to witness some deadliest opponent business houses joining hands to invest all their resources now to promote what they fiercely opposed till yester-years.

* Based upon the above fundaments, the sizes, shapes, layouts, driving inputs, the controls and applications of Velogen units would go with the flight of human imagination. We may now anticipate to fondle with various amazing shapes and sizes of the Velogen Alternators.

* They may look like personal portable C. D. Players or External Hard Drives for computers, or anything equally novel and fascinating, with ultra thin light weight Printed Circuit Boards, like the Compact Disks, having small dots printed on them with some super-conductor, forming the Magnets and Coils, spinning at ultra high speed, driven by the Wind Power if available or even by hand-cranked Coil-Springs.

* Such gadget would be handy, easy and safe to handle by the kids and the elders alike, much more versatile in their functions, serving various individual requirements.

So, let us be ready for the age of Personal or Portable or Palm size Velogen Alternators, just like the Personal Computers, commonly used to generate the required amount of electricity, at any time, anywhere and every where, at absolutely zero cost of production and zero pollution.

MAY GOD BLESS US ALL.

08 CARRY-ON

To present a more clear picture of what is stated above, it is necessary to specifically mentioned here, that all the technical details provided herein are not the outcome of any systematic scientific research or any experiments conducted anywhere.

Even the annexed illustrations are not any scale drawings of any logically approved designs. They are merely the means to carry and deliver some rough image of the basic ideas.

These basic human concepts need to be further carried-on, analyzed, enhanced or modified within the brackets of entire available scientific knowledge and wisdom.

So, please CARRY-ON if you can.

With ALL THE BEST WISHES,

Guru Prem Walia

(ANNEXURE-01)

WIND MILLS in PERSIA

(700 > 1200 A.D.)

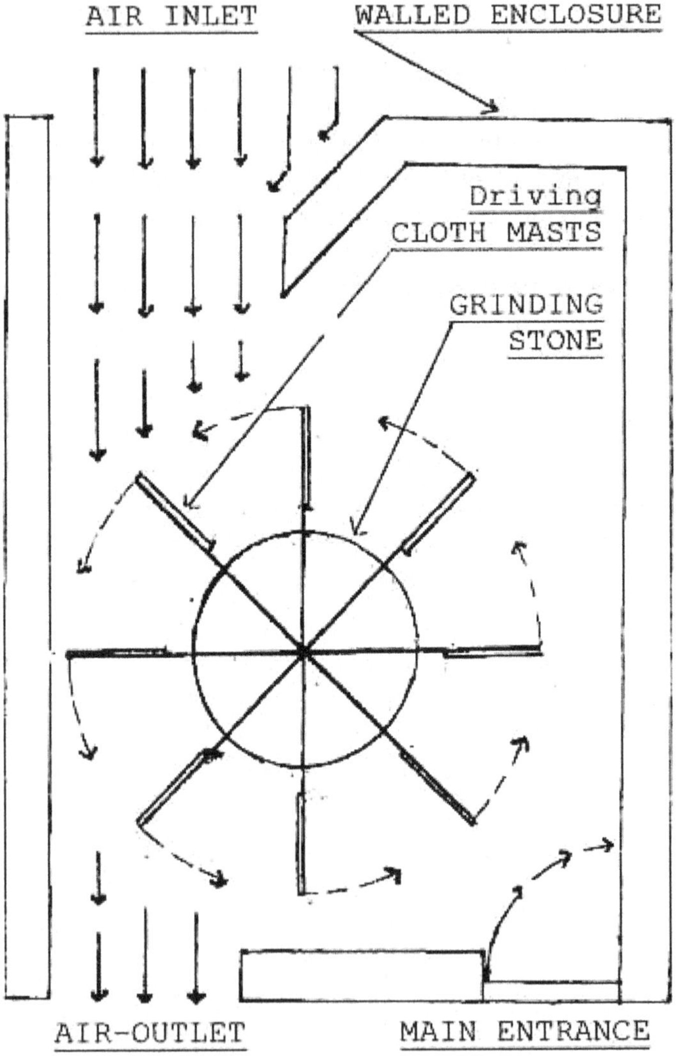

(ANNEXURE-02)

Popular DUTCH WIND MILLS
(1400 A.D.)

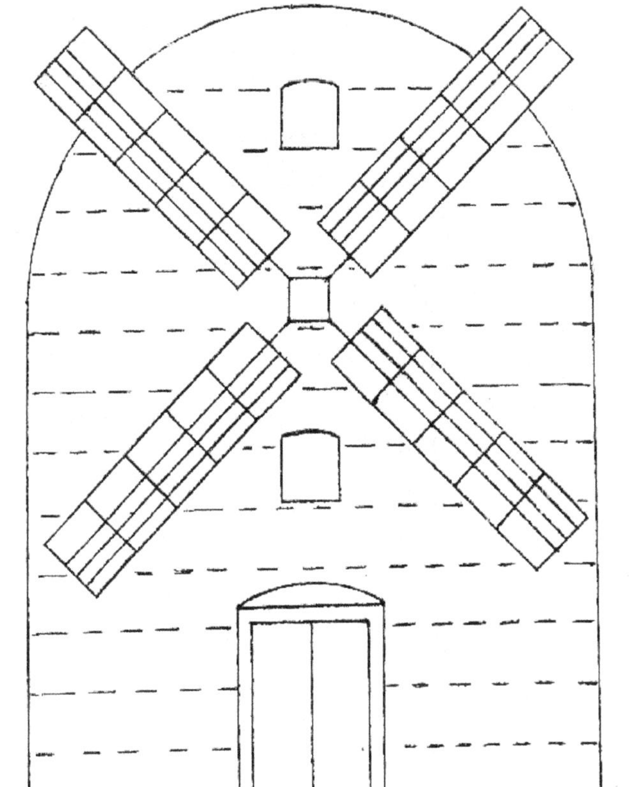

(ANNEXURE-03)

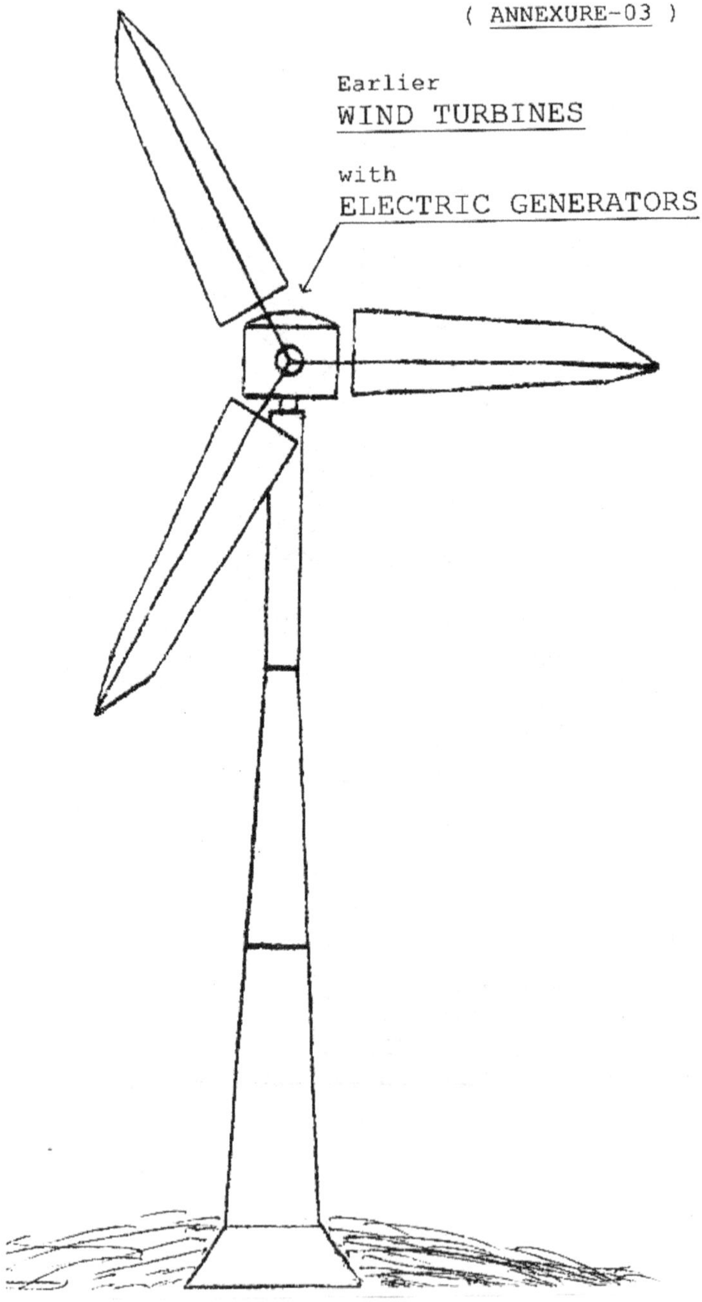

(ANNEXURE-04)

Velogen WIND TURBINE

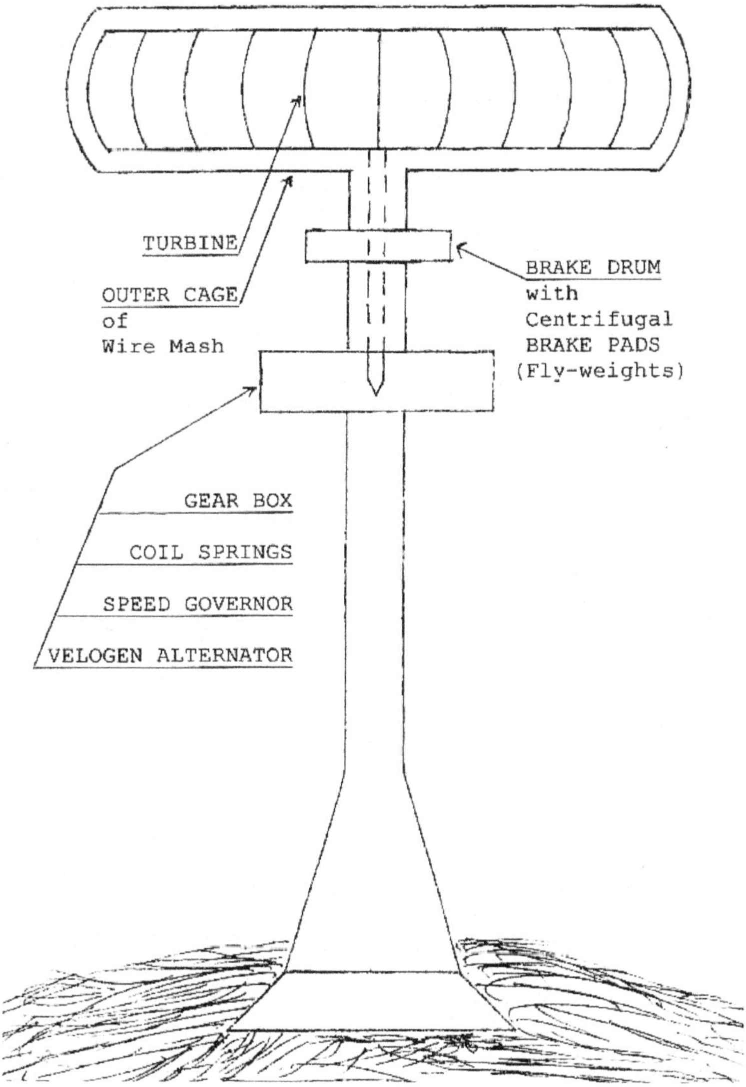

(ANNEXURE-05)

Velogen DISK ROTOR

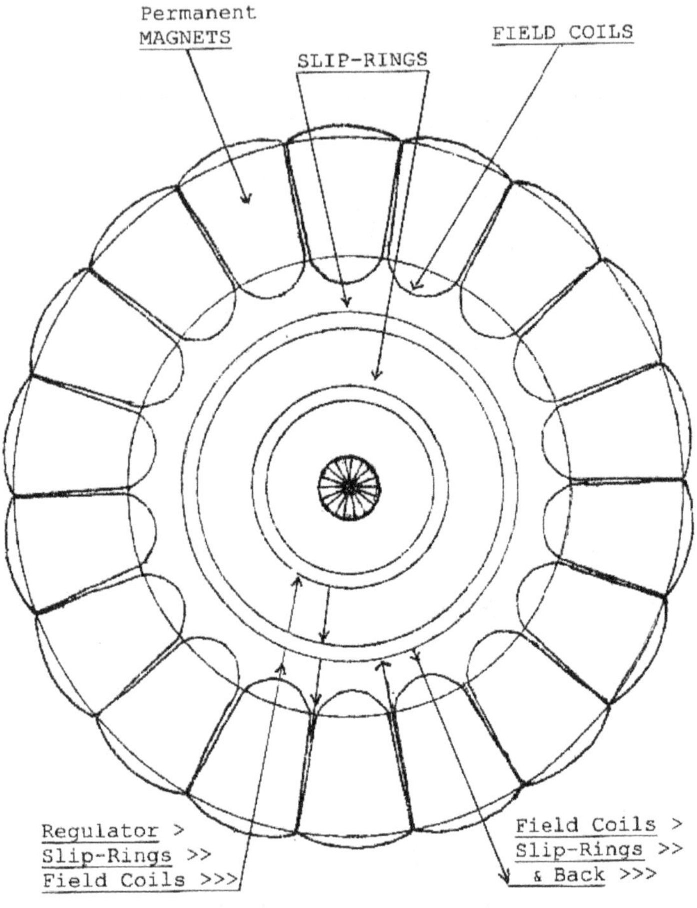

(ANNEXURE-06)

Velogen DISK STATOR

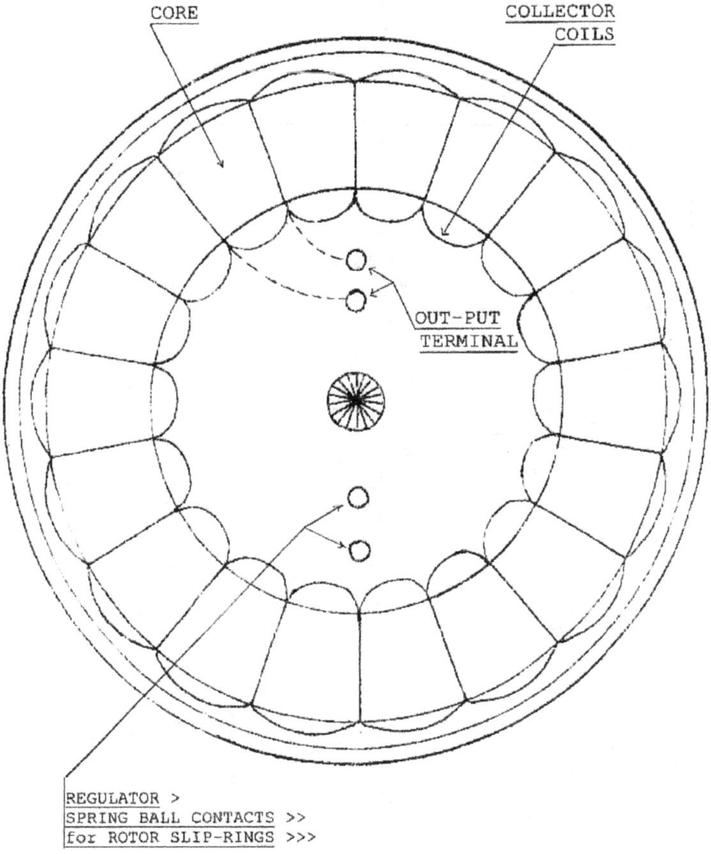

(ANNEXURE-07)

Velogen ALTERNATOR Assembly

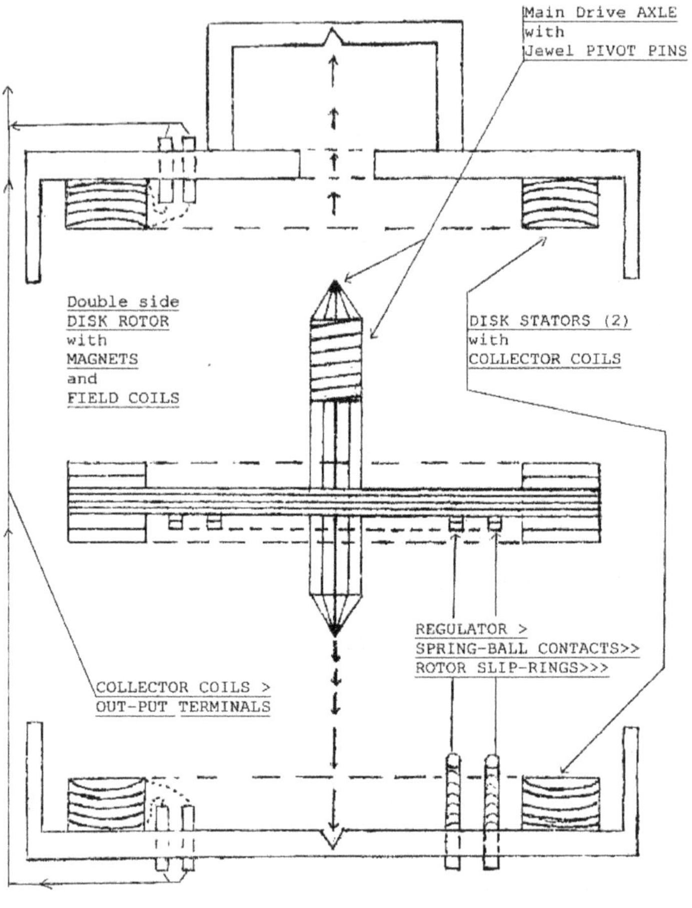

(ANNEXURE-08)

Velogen ALTERNATOR

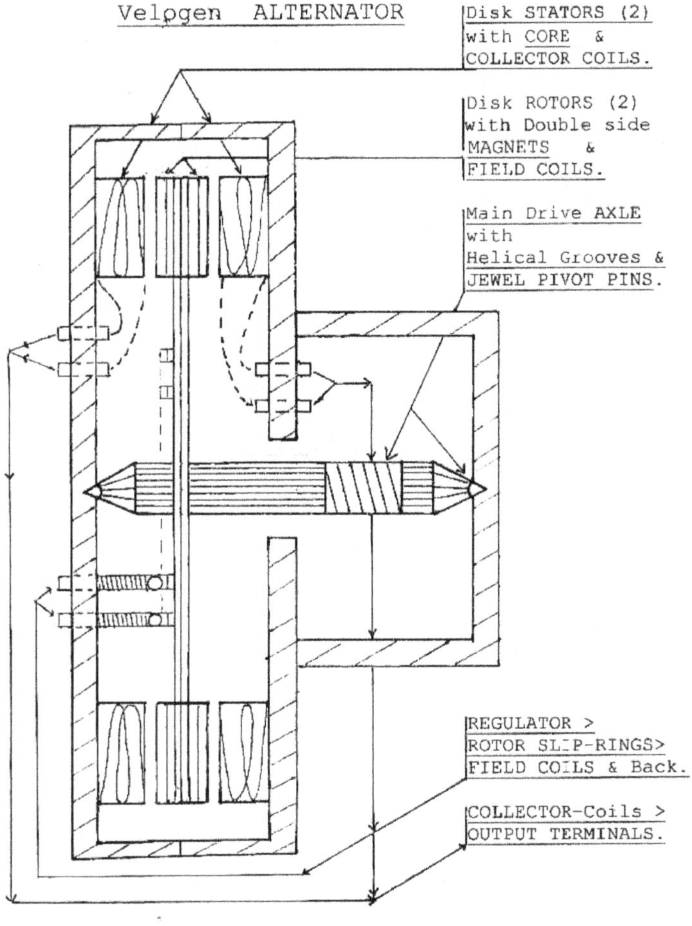

Most valuable REFERENCES

1. CHARCOAL
   http://www.infoplease.com

2. CHARCOAL WARMS THE WHOLE WORLD, Science News Dec. 2001
   http://www.highbeam.com

3. ABOUT FOSSIL FUEL
   http://greenearthsociety.org

4. ABOUT COAL
   http://www.greenearthsociety.org

5. HARMFUL EFFECTS OF COAL BURNING
   http://en.wikipedia.org/Coal

6. ABOUT PETROLEUM AND OIL
   http://greenearthsociety.org

7. ABOUT NATURAL GAS
   http://greenearthsociety.org

8. QATAR OIL
   by Russel Gold,
   Wall Street Journal Feb-2005

9. NATURAL GAS
   Energy Information Administration
   http://www.eia.doe.gov.

10. NATURAL GAS
Interntl. Assn. of Energy Economics,
http://www.iaee.org

11. NATURAL GAS
http://wikipedia.org

12. SYNTHETIC DIESEL FUEL
California Energy Commission,
http://www.ENERGY.CA.GOV/AFVS/
SYNTHETIC DIESEL

13. BIO FUELS
http://www.freeenergynews.com

14. The PROMISE OF HYDROGEN
by Dr. ROBERT J. WILDER
President,
HYDROGEN FUEL CELL INSTITUTE
http://www.h2fuelcells.org

15. HYDROGEN & FUEL CELLS
Three possible approaches to
thinking about the concept.
by Dr. ROBERT J. WILDER
http://www.h2fuelcells.org

16. FIVE TYPES OF FUEL CELLS
by Dr. ROBERT J. WILDER
http://www.h2fuelcells.org

17. FUEL CELL VEHICLES
California Fuel Cell Partnership
http://www.fueleconomy.gov

18. SOLID OXIDE FUEL CELLS
by ULF BOSSEL, June-1998
http://www.efcf.com/reports/

19. DOES HYDROGEN ECONOMY
MAKE SENSE ?
by ULF BOSSEL, April-2005
http://www.efcf.com/reports/

20. ON WAY TO SUSTAINABLE
ENERGY FUTURE
by ULF BOSSEL, Sept. 2005
http://www.efcf.com/reports/

21. WIND TO WHEEL
ENERGY ASSESSMENT
by PATRICK MAZZA
& ROEL HAMMERSCHLAG 2005
http://www.efcf.com/reports/

22. VIABLE ENERGY STRATEGIES,
GROUNDED ON SOURCE
TO SERVICE, ANALYSIS
by JAN H.J.THIJSSEN 2004
http://www.efcf.com/reports/

23. FUTURE OF
THE HYDROGEN ECONOMY,
by ULF BOSSEL, BALDUR ELIASSON
& GORDON TAYLOR
http://www.efcf.com

24. WORLD NEEDS SUSTAINABLE
ENERGY ECONOMY and NOT
A HYDROGEN ECONOMY.
by ULF BOSSEL,
http://www.efcf.com

25. The HYDROGEN ILLUSION
by ULF BOSSEL
http://www.efcf.com

26. FUELS, Source to Service Analysis
by JAN H.J.THIJSSEN
http://www.efcf.com

27. HYDROELECTRICITY
http://en.wikipedia.org

28. TIDAL POWER
http://en.wikipedia.org

29. The ENERGY STORY,
SOLAR ENERGY
http://www.energyquest.ca.gov

30. The HYBRID VEHICLE
http://en.wikipedia.org

31. ILLUSTRATED HISTORY OF
WIND POWER DEVELOPMENT
Part 1 to Part 5
(1000 BC to 2006 and Future))
by Richard Leslie Hills
http://telosnet.com/wind

32. EVOLUTION OF
THE DUTCH WIND MILLS
by MINDY RAN on 17/01/2005
http://www.radionetherlands.nl/
features/dutchhorizons

33. DUTCH WIND MILLS (1891-1976)
by FREDRICK STOKHUYZEN
http:/www.nt.ntnu.no/users/haugwarb

34. POWER FROM WIND,
History of Wind Mill Technology
by Richard Leslie Hills,
Cambridge University Press

35. WIND TURBINES
http://www.wikipedia.org

36. ALTERNATOR SECRETS
http://www.1stconnect.com

37. FUNDAMENTS OF ELECTRICAL ENGINEERING & ELECTRONICS
http://www.vias.org/FEEE

38. ALTERNATORS & GENERATORS THEORY
http://www.rowand.net

39. TURNING WIND POWER ON ITS SIDE
http://www.economist.com/research
March-09, 2006

40 .IMPACT OF WIND FARMS ON BIRDS
http://www.grist.org/advice/2005

41. THE WIND FARM PLANS SCALED BACK
BBC News 04/27/2006
http://news.bbc.co.uk

42. MINISTER EYES VETO ON WIND POWER
by EWIN HANNAN, 04/28/2006
http://www.theaustralian.news.com.

Notes

Notes

Notes

Notes

Made in the USA
Monee, IL
04 November 2026

67385499R00056